SOLITON

The Physics of Sovereignty

Gerard Duffy

UNIFIED POLARITY

Published by Unified Polarity, LLC.

ISBN: 979-8-9936065-1-4

Disclaimer: The concepts and practices discussed in this book are for educational and exploratory purposes. They are not intended as a substitute for professional medical or psychological advice.

"The ocean cannot drown the wave, for the wave is the ocean awake to itself."

Table of Contents:

Prologue

My journey has always been a dual-track exploration. While my formal education was rooted in physics and my professional life spent nearly thirty years in the precise world of scientific instrumentation, a parallel narrative was unfolding. A narrative that would eventually converge into one.

Mystical experiences in childhood fueled a desire to know the mystery of our reality. As I got older, this journey was catalyzed by difficult life challenges and unexplainable moments of synchronicity, which sparked a relentless search for what lies just beneath the surface of the known.

For as long as I can remember, I have felt an undeniable pull toward philosophy, ancient and biblical texts, the deep mysteries of Egypt, and the cosmos itself. I always sensed that the celestial and the earthly were speaking the exact same language, just waiting to be translated.

Eventually, these two tracks merged. I began to see that the rigorous logic of the scientist and the intuitive sight of the mystic were never in conflict. Rather, they are two halves of the same lens. Both are absolutely essential tools for recognizing the invisible ocean we exist within—navigating the fluid dynamics of the ether, understanding the mechanics of reality, and ultimately, reclaiming a lost sense of individual sovereignty.

What This book is NOT

To be clear, this journey isn't for everyone. To maintain the integrity of the message, it is important to state what this book is not:

- This is not a magic pill or a guru-led path. If you are looking for someone to tell you exactly what to think or how to live, you won't find it here. The goal is your own sovereignty, not a new dependency.

- This is not spiritual fluff. We aren't going to ignore the laws of physics or the reality of the material world. We are going to expand them.

- This is not for the dogmatic. If you are firmly committed to a worldview—whether strictly materialist or rigidly religious—that refuses to entertain new data or personal experience, this book will likely be an exercise in frustration.

- This is not a passive experience. Understanding how reality "loads" requires a willingness to look at your own participation in the process. If you prefer to remain a passenger, the truths within these pages may feel too heavy to carry.

Introduction: Soliton: The Physics of Persistence

The Soliton: A Miracle of Balance

In the world of fluid dynamics, there exists a phenomenon that defies our standard expectations of how the world works. Usually, when you drop a stone into a pond, the resulting ripples spread out, lose energy, and eventually vanish into the stillness. This is called dispersion—the natural tendency of things to fall apart.

However, a Soliton is different.

A Soliton is a self-reinforcing wave packet that maintains its shape and velocity as it propagates through a medium. It does not fade, it does not spread, and it does not dissipate. It is the result of a perfect balance between two opposing forces: the force of Dispersion (pulling things apart) and the force of Non-linearity (pulling things together).

In the language of physics, a Soliton is an invincible pattern of information. It doesn't fight the fluid; it utilizes its very pressure to hold itself together.

But the Soliton carries a meaning far beyond the water. It is a living metaphor for the part of us that longs to remain intact while moving through the chaos of the world.

This is where physics begins to speak directly to the human experience. A life, like a wave, is constantly moving through forces that try to pull it apart: fear, grief, pressure, distraction, uncertainty, and the endless noise of the external world. Without an internal organizing principle, the self begins to disperse. We lose shape. We become reactive. We are carried by whatever force is moving through the field at the time.

A Soliton reveals another possibility. It shows that persistence is not achieved by resisting the medium, but by becoming coherent within it.

When we look at our lives through this lens, we realize we have been operating from a fundamentally different — and much more fragile — model of existence.

If a Soliton represents a state of total integration and sovereignty, most of us are currently living in a state of fragmentation. We are not moving through life as coherent waves. We are reacting to life as separate objects.

To understand how this fragmentation manifests, we can look at a fluid medium we all recognize: the ocean. When we misidentify our true nature, we lose our internal resonance and become subject to the turbulence. We stop being the wave that moves through the ocean and start being the debris moved by it.

The Illusion of the Cork

Most humans do not perceive themselves as Solitons. Instead, they perceive themselves as Corks.

We feel like isolated objects bobbing on the surface of a vast, chaotic reality. The problem is that a Cork can never truly feel safe because it believes it is separate from the ocean.

Through years of inner work and contemplation, I have come to realize that this is a limited perception of our reality.

We are not corks bobbing on the ocean; we are an expression of the ocean itself. We are not separate from the fluid; we are the fluid in a specific, coherent state of motion. This shift in perspective—from the separate object to the integrated wave—is the key to reclaiming your power.

The Witness and the Fluid

My background is in physics—the study of how energy and matter interact to create the experience we call "reality". My years of applied introspection led me to the understanding that the separation we feel as Corks is a trick of the senses.

By lowering the internal noise through silence and presence, I began to perceive the universe not as a collection of static objects, but as a high-pressure, super-fluid medium—the Aether. In these deep states of observation, I have witnessed the fluid dynamics of existence directly.

This book is the manual for that transition. It is the journey from being a Cork—exhausted by the struggle to survive—to being a Soliton—an integrated being that understands how to use the viscosity and friction of life to lock into a state of absolute sovereignty.

We are going to move past the illusion of separation and into the truth of expression. We are going to look through the eyes of a mystic and a physicist to see the engine running behind the surface. And in doing so, we will realize a truth that changes everything:

Don't fight the water. Become the wave.

Chapter 1: The Structural Fracture

Awakening and the Source Code of Meaning

The Early Receivers

Before sovereignty becomes a conscious path, it often begins as a disturbance—a subtle frequency felt in the background long before the signal grows loud enough for the rest of the world to notice.

For some, this tuning is present from an early age. Those with highly sensitive nervous systems—what society sometimes categorizes as Highly Sensitive People, or HSPs—often experience the world with the gain turned all the way up. They feel the overwhelming weight of the physical senses: the frustration of unnecessary noise, the agitation of crowded spaces, and the chaos of modern life.

But this heightened state is not only a matter of sensory overwhelm. Beneath the physical intensity, there is often another kind of perception beginning to stir—faint, unnamed, and difficult to explain. It is not yet a clear message, and it may not yet have a name. It is simply the subtle sense that something else is present beneath the surface of things, something deeper than the obvious world everyone has agreed to call real.

At first, this background signal may appear only as curiosity, sensitivity, unease, or a strange refusal to accept the surface explanation. But for those who feel it early, it becomes an invitation to explore the hidden architecture of reality. These individuals often act as canaries in the coal mine, sensing the undercurrents of a shifting paradigm before it breaches the collective surface.

This capacity to feel beyond the veil is not reserved for a specific neurotype. The signal is now broadcasting to everyone. Humanity is standing at a collective threshold, and what was once a quiet disturbance is becoming impossible to ignore.

The Hard Reboot

For many, the awakening of the senses is catalyzed by a hard reboot—a severe personal crisis such as a sudden health issue, a profound emotional loss, deep disillusionment, or a Dark Night of the Soul. While some experience a gradual opening, the intense pressure of a personal trial is often what strips away the surface noise and forces the internal signal into clarity. At first, this crisis appears only as suffering; the external world becomes unreliable, old explanations stop working, and the life you carefully assembled feels strangely insufficient, as though something deeper is pressing against the walls of your perception. Forced to turn inward through presence, silence, and the sheer exhaustion of trying to control the uncontrollable, a subtle shift occurs. Over weeks, months, or years, your perception of reality undergoes a distortion. The world you took for granted becomes far less absolute, and the concrete architecture of your life begins to lose its density. This shattering of your beliefs about security and control can be deeply unsettling, yet within that crumbling, something else is revealed: you are not merely a separate observer moving through an external world, but a participant within a living field of intelligence..

As your awareness aligns with this field, the background static of the world begins to fall away. You start to notice patterns that previously felt like random noise. You see 11:11 on a digital clock, 333 on a license plate, or the exact word you were just thinking about appears on a billboard at the moment it crosses your mind. Initially, these moments feel like a wink from the universe—a small sign that you are not alone.

As the internal resistance begins to drop, the experience can become even more profound and, at times, overwhelming. You might think of a specific name, and suddenly that name is everywhere you look—on signs, in books, and in the mouths of strangers. It starts to feel as if the world is no longer a solid, indifferent place, but a responsive mirror that is listening to your very thoughts.

The Trap of the Start Line

There is a common misconception that this initial "Awakening" is the destination. Often, when people feel that first spark or start seeing these patterns, they believe they have reached the finish line. They may claim enlightenment or believe they have "figured it out." But in the physics of the soul, Awakening is not the destination; it is merely the start line.

If Awakening is the spark that lights the fire, Sovereignty is the discipline required to keep that fire burning without being consumed by it. True sovereignty is a muscle that must be built through the real, often painful work of holding your center between the chaos of the world and the intensity of your own emerging truth. When you mistake the start for the finish, you are often slammed back down by the very reality you thought you had mastered. This is where the real work begins. It is the move from knowing about the code to living as the code.

The Resonance Squeeze

As you begin your journey into sovereignty and your internal rhythm changes, the world you once knew begins to feel strangely unfamiliar. This is the distortion—a stage where the social and emotional gravity that once held your life together begins to loosen.

This is a time of intense emotional volatility. You may find yourself experiencing waves of deep clarity, exhilaration, and lightness, followed by days of unexpected tiredness, sadness and confusion. It is not a punishment, nor is it a sign that you are doing something wrong; you are simply being pulled through a transitional space. Like shedding a layer of clothing that no longer fits, you are letting go of the cork-state that fought to survive. For me, it felt like I was being pulled into a tightening funnel, a vortex, stripping away layers of identity.

One of the most delicate parts of this transition is the natural evolution in friendships. As your frequency shifts, you find that you no longer resonate with certain environments or people you may have known for years. This isn't a matter of judgment; it is a matter of resonance. When the frequency match is gone, the connection often fades, leading to a temporary sense of loneliness. You are grieving not just for the people you are leaving behind, but for the version of yourself that fit so neatly into their lives.

Physical Recalibration: The Frequency Upgrade

This transition has a physical aspect as well. As your internal signal shifts, your vessel undergoes a period of recalibration to handle the new bandwidth. This can manifest in several ways, such as fatigue, unexplained aches, back pain, or digestive issues. Ringing in the ears is another common symptom and this ringing, which can vary in pitch and intensity, is often a sign of a frequency upgrade—your nervous system attempting to match the higher resonance, akin to tuning a radio dial to a new station. This is not a sign of sickness, but a sign of your hardware being updated to support the new operating system.

Stepping Into the Fire

Ultimately, this stage of the journey is about having the willingness to accept the reality that is emerging right in front of you, rather than clinging to the familiarity of the past.

Becoming a Soliton—a coherent, self-sustaining wave—requires you to do the inner work of integrating your shadows and honoring your own truth, even when it feels vulnerable to do so. You are no longer just a passenger being tossed by the waves; you are learning to participate in a grand, responsive dialogue with the Field around you.

Chapter 2: The Vessel

The Biology of the Machine

The Architecture of the "Fixed" Life

If you were to look at a medical textbook, you would see the human body described as a collection of distinct systems. You have the skeletal system for structure, the muscular system for movement, and the nervous system for communication. We are taught to view ourselves through a compartmentalized lens—a series of separate organs—hearts, lungs, livers—working together like the parts of an engine.

In this "biological filter," we are essentially slaves to the system. We believe that if an organ breaks, we are broken. We believe that if our tissues show signs of age or disease, we are simply experiencing the random noise of a biological machine winding down. Because the system is compartmentalized, we feel fragmented. We don't see the unified field that connects the heart to the mind; we see two different doctors for two different problems. We have become biological slaves to a machine we were never taught to operate.

When we live through this filter, we become the Cork.

The Passive Victim and the Frantic Fixer

In the introduction, we touched on the image of a cork bobbing in the ocean. At first glance, a cork seems benign, even peaceful. But look closer. A cork is a consequence of its environment. It has no engine. It has no steering. It is being beaten around by the waves, reacting to every external shift. If the water is cold, the cork is cold. If the water is turbulent, the cork is tossed.

The cork represents a state of total passivity. But human beings are rarely content to just bob. Because the cork-state is one of zero agency, it eventually gives birth to a deep, underlying fear. When you feel you have no power over your own vessel, the ocean becomes a threat.

This is where the Trap of the "Fixer" begins.

Because the cork feels victimized by the water, its primary instinct is to try to manipulate the environment through sheer force. We go around trying to "fix" things. We try to fix our partners, our careers, and our bank accounts. In our bodies, if we feel a symptom, we try to suppress it or fix it with an external intervention.

This is a reactive, often aggressive masculine resistance. We are trying to fix the ocean so we can finally be still. We think, "If I can just make the water stop moving, I will be safe." But a cork can never dominate the ocean. You are a fixer who is perpetually exhausted because you are fighting the physics of the environment rather than mastering the physics of your own vessel. You are trying to use a localized, frantic effort to solve a systemic, universal problem.

The Symphony of Systems

To understand why "fixing" the parts fails, we need a better metaphor: The Orchestra.

Imagine your heart, your lungs, your liver, and your nervous system are not just parts in a machine, but world-class musicians sitting in a concert hall. Each organ has its own unique sound—the heart has its rhythmic pulse, the lungs have their airy swell, and the nervous system provides the intricate, high-speed melodies of data transmission.

When these systems are in balance, you experience health as a grand symphony. It is a state of Coherence, where every instrument is in tune and playing from the same sheet music.

However, in the Cork state, the musicians lose their way. The heart starts to beat in a frantic, erratic rhythm; the lungs become shallow and tight. Without a central organizing principle, the orchestra descends into chaos. This is what we call illness or inflammation. It isn't necessarily that the musicians (the organs) are broken. It is simply that they have lost their connection to the conductor.

The Soliton as Conductor

In a traditional orchestra, the conductor doesn't play an instrument. Their job is to provide the Signal. They set the tempo and the unity of the performance.

This is exactly what the Soliton does for your biology. A Soliton is a wave that refuses to disperse. It holds its integrity through the sheer power of its internal focus. When you learn to generate this state of internal coherence, you are essentially stepping onto the conductor's podium. You are providing a singular, powerful rhythm that every cell in your body can hear.

Instead of being the fixer frantically trying to tell the violin player to play better, you improve the signal. When the conductor's baton moves with clarity and strength, the musicians naturally fall back into tune. The noise resolves back into music.

Signal vs. Noise: The Spectrometry of Life

In my professional career, I have spent years working with mass spectrometers—sophisticated instruments designed to identify the exact chemical makeup of a substance by analyzing its mass-to-charge ratio. Whether analyzing atmospheric gases or complex surfaces, a fundamental challenge in the laboratory is Signal-to-Noise ratio.

"Noise" is the background interference—the random, unwanted data that obscures the truth. The "Signal" is the specific, coherent information we are looking for. In any scientific instrument, data quality is defined by its signal-to-noise ratio. When a true signal is no longer observable, it is usually because the data is simply buried in the noise. The system has become contaminated with the residue of previous samples, creating a background static so high that the actual signal cannot break through. Your biology works on the exact same principle.

Most people are living with a biological Signal-to-Noise ratio that is dangerously skewed toward the Noise. The Noise is your stress, your reactive mentality, and your fragmented view of your own body. This noise drowns out the Signal—the natural blueprint for health and sovereignty. To become a Soliton, you must learn to dampen the noise and amplify the signal. You must become the operator of your own internal spectrometer, cleaning the noise of your mind until the background static of the world no longer obscures your truth.

The Ghost of the Machine Age

Before we can truly understand our potential, we have to recognize that we are fighting against two hundred years of bad metaphors. During the Industrial Revolution, the world became obsessed with the steam engine. We inherited the idea that we are a collection of parts that eventually wear out and must be fixed by an outside mechanic.

But you are not a steam engine. A steam engine is entropic—it starts dying the moment it is built. You, however, are a biological miracle that has the ability to self-organize, self-repair, and evolve in real-time. You are a Symphony of Information.

The High-Tech Librarian: DNA as Code

In the mid-1950s, Francis Crick realized that DNA was not just a chemical; it was Information. This was the moment biology merged with Information Technology. If our tissues and organs are the "hardware," then DNA is the "source code."

Think of your DNA as a massive library. Within every cell, you have a library containing every book (gene) ever written by your ancestors. Just because a book is sitting on the shelf does not mean it is being read. This is the science of Epigenetics.

In this metaphor, you are the Librarian. Your internal state—the "Signal"—is what determines which book you pull off the shelf today.

- If you are in a state of chronic stress (The Cork), the Librarian pulls down the "Survival and Defense" manuals.
- If you are in a state of coherence (The Soliton), the Librarian reaches for the "Repair and Vitality" volumes.

The books (the genes) don't change, but the story of your life changes based on which books you choose to open. You are not a victim of the library; you are the one holding the library card.

The 98%: The Dark Matter of the Self

Only about 2% of our DNA actually codes for proteins. The other 98% was once dismissed as "Junk DNA." But as we move into the Age of Aquarius—an era defined by networks, frequency, and the collapse of old hierarchies—we are realizing this "junk" is actually a Biological Antenna Array.

These non-coding regions are the "Dark Matter" of your genome. They act as the sensors and routers of your biology. They are constantly receiving signals from the "Field" around you and coordinating the response of every cell. When you achieve the Soliton State, you are essentially upgrading your network protocol. You move from a "dial-up" connection of fear to a fiber-optic connection of coherence.

Feeding on the Sun: Schrödinger and Negentropy

In 1944, physicist Erwin Schrödinger argued that life "feeds on negative entropy." According to the Second Law of Thermodynamics, the universe moves toward disorder (Entropy), defined by Boltzmann's equation:

$$S = k \ln W$$

Life, however, exhibits Negentropy. It takes noise and turns it into order. A seed takes the random elements of dirt and light and organizes them into a flower.
The Cork is entropic: It is winding down, letting the environment dictate its decay.
The Soliton is negentropic: It is a localized pocket of perfect structure moving through a sea of chaos.
When you are in a Soliton state, you aren't just trying not to die; you are actively organizing your own biology through the power of your own signal.

But achieving this Soliton state requires a profound internal mechanism: the integration of two opposite forces within the psyche. In fluid dynamics, a Soliton only forms when the force trying to pull a wave apart is perfectly balanced by the force trying to crash it together.

In the human operating system, this means holding the tension between your most reactive, fear-driven instincts and your uncompromising structural integrity. When you stop allowing fear to trigger an automatic reaction, you can actually harness its intensity, using it to reinforce your own coherent structure.

Once you integrate these opposites, you stop being tossed by the chaotic fluid of the world. You take control of your own emotional and energetic output. You move from the guest user to the system administrator.

Conclusion

You are not a collection of separate organs destined for decay. You are an Orchestra waiting for a Conductor, a Library waiting for a Librarian, and a Spectrometer waiting for the noise to be dampened.

The question is no longer "How do I fix my broken parts?"

The question is: "How do I become the Conductor?"

Chapter 3: The Soliton

The Physics of Identity

The Great Wave of Translation

In August of 1834, a young Scottish engineer named John Scott Russell was observing a boat being drawn along a narrow channel by a pair of horses. When the boat suddenly stopped, something extraordinary happened. Instead of the water simply churning and dying out, a mass of water rolled forward from the bow in the shape of a large solitary elevation—a rounded, smooth, and well-defined heap of water that did not dissipate..

Russell, sensing he was witnessing something that defied the known laws of physics, jumped on his horse and followed it. He chased this single wave for over two miles, watching in disbelief as it maintained its height, its shape, and its speed without ever fading. He called it the "Wave of Translation."

At the time, the scientific community laughed at him. According to the linear mathematics of the day, such a wave was impossible. Waves were supposed to disperse; they were supposed to flatten and disappear as they moved. But Russell had seen the truth: a wave could, under the right conditions, become a Self-Sustaining Entity. He had discovered the Soliton.

The Invisible Ocean: The Aether as Fluid

Before we go further into the nature of this wave, we must address a common misunderstanding about the world we live in. When we use words like "ocean," "waves," and "flow," it is easy to assume we are speaking metaphorically about water. But to a physicist, these are not metaphors; they are descriptions of Fluid Dynamics.

Most people think of a fluid as something you can drink—water, juice, or coffee. But in the language of science, a fluid is simply any substance that flows and deforms under pressure. This includes liquids, yes, but it also includes the air you are breathing right now. It includes the plasma of the sun. And, as we are beginning to rediscover, it includes the very "empty space" between the stars.

For centuries, the greatest minds in science believed the universe was filled with an invisible, highly subtle fluid called the Aether (or Ether). They realized that just as sound needs air to travel, light and electromagnetism must need a "medium" to move through. Modern Quantum Field Theory has brought this back under the name The Zero-Point Field. We now know that "empty space" is not empty at all. It is a dense, fluid-like medium of energy that behaves exactly like water, only at a much finer scale.

The laws of Fluid Dynamics are universal—they scale from the swirl in your coffee cup to the rotation of a galaxy. The reason this is vital to your sovereignty is simple: A Soliton is a wave that can travel through any fluid medium. Whether it is water, air, or the Aether of the universe, the physics remains the same. You are a wave traveling through the most sophisticated fluid in existence.

Non-Linearity: The Secret of the Shape

Why did that wave stay together while every other wave in the channel fell apart? The answer lies in a concept that is the foundation of your sovereignty: Non-Linearity.

In a linear world, 1 + 1 always equals 2. If you hit a wave with a certain amount of pressure, it displaces a certain amount of water.

But in a non-linear system, the output is not proportional to the input. The system begins to interact with itself. In the case of the Soliton, two opposing forces of nature suddenly start working together in a perfect handshake:

1. Dispersion: The tendency of a wave to spread out and lose energy (The flattening force).

2. Non-Linearity: The tendency of the medium (the fluid) to "push back" and sharpen the wave (The structuring force).

When these two forces reach a state of perfect balance, they cancel each other out. The wave "locks." It becomes a localized packet of energy that no longer needs an external push to keep going. It becomes its own cause.

The Identity Equation

In modern physics, we describe this phenomenon using the Non-Linear Schrödinger Equation (NLSE). While it looks complex, it tells a very simple story about who you are:

$$i\frac{\partial\psi}{\partial t}+\frac{\partial^2\psi}{\partial x^2}+k\,|\,\psi\,|^2\,\psi=0$$

In this equation, ψ (psi) represents the wave—the "You." The second term ($\frac{\partial^2 \psi}{\partial x^2}$) is the dispersion—the noise of the world trying to pull you apart. The third term ($\kappa |\psi|^2 \psi$) is the non-linearity—your internal focus, intensity, and coherence.

When your internal intensity ($|\psi|^2$) is high enough, it perfectly balances the external dispersion. You stop being a person who is falling apart and start being a Coherent Identity. This is the mathematical proof of sovereignty: you are the only entity in the universe capable of balancing the chaos of the world with the intensity of your own signal.

You are not the Atoms

This leads us to a profound realization: You are not your body.

If you look at your body through a microscope, you will see that you are made of roughly 7 octillion atoms. But those atoms are constantly leaving. You breathe them out, sweat them out, and replace them through the food you eat. Every seven years, almost every single atom in your body has been replaced.

The truth is that you are not the water; you are the

pattern moving through the water. You are the Soliton that organizes those atoms into a shape called "You." Your vessel is the medium, but your Identity is the Information. When you realize this, the fear of breaking disappears. The water can be turbulent, the atoms can be replaced, but the Geometry remains.

Mathematical Invincibility

One of the most startling properties of Solitons is what happens when they collide. When two normal waves hit each other, they create a mess; they interfere, they distort, and they often destroy each other's shape.

But when two Solitons collide, they pass right through each other as if the other didn't exist. They might experience a slight phase shift (a momentary change in timing), but they emerge on the other side with their shape and velocity completely intact.

This is the goal of the Sovereign Being. Most people are Linear Waves—when they hit a conflict, a toxic person, or a global crisis, they lose their shape. But when you operate as a Soliton, you can move through the interference patterns of life without losing your integrity. You don't need to fight the other wave; you simply need to hold your own frequency so perfectly that the collision cannot "stick" to you.

The Universal Scale of Flow: As Above, So Below

One of the most beautiful aspects of fluid dynamics is its scale-invariance. This is a scientific way of saying that the universe is "fractal"—the patterns that govern the very small are the exact same patterns that govern the very large.

In philosophy, this concept has been known for millennia as "As Above, So Below." While ancient seekers used this phrase to describe the mystical correspondence between the macrocosm of the stars and the microcosm of the soul, modern physics has finally provided the mechanical proof. The universe doesn't have different sets of rules for different sizes; it has one Master Code that scales.

The only thing that changes is the density and the viscosity of the fluid:

- Water is a dense, thick fluid. It moves slowly, and we can feel its weight.
- Air is a less dense fluid. We can move through it easily, but when it moves fast, we feel its power.

- The Aether is a "super-fluid." It is incredibly subtle and moves at the speed of light, yet it is the foundation upon which all matter is built.

The Soliton is the universal Master Pattern that works in all three. Whether you are John Scott Russell's wave in a Scottish canal or a Sovereign Being holding your shape in a chaotic social environment, you are utilizing the same fluid laws. You are organizing the medium—whatever its density—into a coherent structure.

As above, so below; as within, so without. The physics of the galaxy and the physics of your heart are playing the same song.

A Note on the Horizon: The Persistence of the Pattern

If a Soliton is, by definition, a wave that holds its shape across different mediums, we must eventually ask the ultimate question: What happens when the medium itself—the physical vessel—reaches its limit?

If you are not the atoms, but the information that organizes them, then the transition we call "death" takes on a completely different geometry. We will explore the profound implications of this persistence in later chapters. For now, simply consider this: a signal does not cease to exist just because the radio is turned off. It simply returns to the field, waiting for the next frequency to catch it.

Chapter 4: The Ocean

The High-Pressure Field of Reality

The Fish and the Water

There is an old story of two young fish swimming along who happen to meet an older fish swimming the other way. The older fish nods at them and says, "Morning, boys. How's the water?" The two young fish swim on for a bit, and then eventually, one of them looks over at the other and goes, "What the hell is water?"

This is the state of the average human being. We are so immersed in the medium of our existence that we have become completely blind to it. We look at the space between our hands and call it "empty." We look at the vast reaches between the planets and call it a "vacuum."

But as we established in the previous chapter, there is no such thing as "nothing." The universe is not a series of objects sitting in a void; it is a single, continuous, high-pressure fluid. To understand the Soliton, you must first understand the Ocean.

The Lie of the Empty Room

Our senses are evolved for survival, not for deep physics. To our eyes, air is invisible, and space is a "hole" where things aren't. We have been taught to view the universe as a container filled with a few bits of matter—stars, planets, and people—rattling around like marbles in a box.

In the old paradigm, we believed that if you took all the stuff out of a room, the room would be empty. But modern physics tells a different story. If you were to remove every atom from a cubic centimeter of space, you wouldn't be left with nothing. You would be left with the Zero-Point Field—a seething, high-energy foam of potential.

Space is not a void; it is a plenum. It is the most solid thing in existence, even if our eyes can't register the frequency of its density.

The Lightbulb and the Boiling Sea

To grasp the sheer amount of energy we are talking about, we have to look at Energy Density. The late physicist David Bohm and others calculated the background energy of the vacuum. Their findings were staggering: the energy contained in a single cubic centimeter of empty space (about the size of a sugar cube) is greater than the energy of all the matter in the known universe.

Think of it this way: if you could tap the energy in the space of a single lightbulb, you would have enough power to boil all the oceans on Earth.

Why don't we feel this? Because the pressure is perfectly equalized. Just as a deep-sea fish isn't crushed by the thousands of pounds of water pressure above it—because that same pressure is inside its cells pushing back—we don't feel the "weight" of the Aether because we are made of it. We are Fluid beings living in a Fluid world.

The Great Shift: From the Ice of Pisces to the Flow of Aquarius

The reason we are only now beginning to "see the water" is because we are undergoing a massive transition in human consciousness—a shift between two distinct Ages within the Great Year.

For the last 2,000 years, we have lived in the Age of Pisces. In the physics of the Great Year, this era represented an unconscious phase of the cycle; a deep winter. In this phase, the Aetheric medium was at its most viscous—essentially a state where the flow of information was so restricted it became frozen. This was the age of structural rigidity. Because the medium offered so much resistance to the signal of connection, we could only perceive the world through the lens of separation. It was the "age of objects," where we viewed ourselves as isolated entities (corks) trying to survive in a cold, indifferent, and mostly empty universe. This paradigm gave us the Industrial Revolution, the "machine" metaphor for the body, and a deep sense of existential loneliness.

Now, we are entering the Age of Aquarius. The symbol for Aquarius is the Water Bearer, but she is not just carrying water; she is pouring it out. In the Aquarian Age, the Ice is melting. We are moving from the "age of objects" to the "age of the field." We are realizing that the "water" she pours is the Aether itself—the medium that connects all things. This era is defined by frequency, information, and the collapse of distance. In this new world, you aren't an object in a void; you are a Wave in the Field. To survive and thrive in this age, you have to stop looking at the "fish" and start understanding the "water."

Navier-Stokes and the Architecture of Turbulence

When people think of "The Field" or "The Aether," they often imagine something static and calm. But fluids are rarely static. They move, they swirl, and they become turbulent. In fluid dynamics, we use the Navier-Stokes equations to describe this motion:

$$\rho \left(\frac{\partial \mathbf{u}}{\partial t} + \mathbf{u} \cdot \nabla \mathbf{u} \right) = - \nabla p + \mu \nabla^2 \mathbf{u} + \mathbf{f}$$

While this looks like a complex array of symbols, it describes a fundamental truth: Chaos is structured. When water moves through a pipe or air moves over a wing, it creates vortices and eddies. Most people see turbulence in their lives—stress, conflict, sudden changes—as a mistake. They see chaos as a sign that something is wrong.

But look at the equation. Turbulence is a natural result of Velocity and Pressure. Without turbulence, there is no resistance. And without resistance, a Soliton cannot form. A Soliton needs the non-linear "push-back" of the medium to hold its shape. In other words: The chaos of the ocean is the very thing that keeps the wave from falling apart.

Pressure as Support

This is the fundamental shift from the Cork to the Sovereign.

- The Cork feels the pressure of the ocean as a threat. It feels beaten around by the waves of life.
- The Soliton understands that the pressure is its Support.

Imagine a balloon. If there were no air pressure on the outside, the balloon would simply expand until it popped. It is the atmospheric pressure squeezing the balloon that gives it its round, coherent shape.

Your life is exactly the same. The challenges you face, the "weight" of the world, and the high-pressure environment of the Age of Aquarius are not there to crush you. They are there to define you. They are the Non-Linear forces that allow your internal signal to lock into a Soliton. When you stop resisting the pressure and start utilizing it, you move from being a victim of the environment to being a master of the flow.

Conclusion: The Etheric Sea

The Ocean is not a void. It is a high-pressure, super-fluid medium of infinite potential. It is the Mother of the wave. You are not a lonely object drifting in a vacuum; you are a vibrant, coherent pattern moving through a sea of energy that is literally supporting your existence with every squeeze of its pressure.

Chapter 5: The Signal

The Art of Unified Polarity

The Internal Oscillator

We have established that you are a Vessel (the hardware) and a Soliton (the software) moving through an Ocean (the field). But what generates the wave? In physics, every wave requires a source—an oscillator.

In your biology, you have two primary oscillators: the Brain and the Heart.

The Brain is your electrical transmitter. It processes data, analyzes patterns, and creates the high-frequency buzz of your thoughts.

The Heart is your magnetic transmitter. It produces an electromagnetic field that is up to 5,000 times stronger magnetically than the brain, extending several feet beyond your body in a toroidal shape.

When these two oscillators are out of sync—when you are acting one way but feeling another—your Signal is scattered. You are broadcasting Noise. This is the state of the Cork: a fragmented broadcast that the ocean simply ignores. But when your thoughts (Brain) and your emotions (Heart) align, they achieve a state called Coherence. In physics, this is known as Phase Locking.

Your signal becomes a singular, focused beam of information. It transitions from a weak, flickering candle to a coherent laser. This laser is the only thing powerful enough to organize the high-pressure Aether into a new reality.

The Law of Two Poles: Scale Invariance of Polarity

To understand how this signal becomes a Soliton, we have to look at one of the most fundamental laws of the universe: Polarity. As we discussed in the concept of As Above, So Below, the rules of the universe don't change as you move through different scales.

Look at the Earth: It has a North Pole and a South Pole. Without this polarity, the Earth would have no magnetosphere, meaning it would have no protection from solar radiation. Look at a Magnet: You can cut a magnet into a thousand tiny pieces, and every single piece will still have a North and a South Pole.

This is Unified Polarity. The North isn't better than the South. They are not enemies; they are two ends of the same stick. The tension between them—the "Push and Pull"—is exactly what creates the Field. Most people spend their lives trying to be "one-pole" beings. They want the light without the dark, the signal without the friction. But in the physics of a Soliton, if you remove one of the poles, the wave collapses into a flat, stagnant line.

The Shadow: The Force of Dispersion

This brings us to the work of Carl Jung and the concept of the Shadow. Jung realized that every human being has a "Shadow"—a collection of traits, fears, and primal energies that we deem negative and attempt to suppress.

In the paradigm of the Fixer, we view the Shadow as a broken part of ourselves that needs to be removed. But let's look at the math of the Soliton again:

$$i\frac{\partial \psi}{\partial t} + \frac{\partial^2 \psi}{\partial x^2} + |\psi|^2 \psi = 0$$

Remember that second term? $\frac{\partial^2 \psi}{\partial x^2}$ represents Dispersion. This is the mathematical tendency of any wave to spread out, lose energy, and vanish. In your psyche, the Shadow is the force of Dispersion. It is the doubt, the hidden fear, and the "dark" energy that threatens to pull your identity apart.

The mistake most people make is trying to fix the Dispersion by ignoring it. But in a Soliton, the intensity of the signal (the third term) only locks because it is perfectly countering the Dispersion. Without the Shadow, there is no push-back. Without the push-back, the signal has nothing to grip onto.

The Great Thaw: The Inevitability of Integration

As we discussed in the previous chapter with the concept of the Great Year, the phase of time we are moving through dictates the viscosity of our collective reality. For the last 2,000 years of the Age of Pisces, the Aetheric medium was in a state of high viscosity—what we can call the Ice state.

In this rigid, frozen environment, it was easy to suppress the Shadow. We could bury our secrets and our traumas deep in the permafrost of the subconscious. Because the medium was so rigid, those clogs stayed in place, and we could survive without looking at them.

But the Ice is melting. As we move into the Age of Aquarius, the frequency of the field is rising and the medium is becoming fluid. This creates a thermodynamic necessity: The Great Thaw. Everything that was previously frozen and hidden is now being forced into the current.

The Shadow is not surfacing to punish you; it is surfacing because the medium can no longer hold the weight of the secrets. Just as a thawing river pushes all the debris to the surface, this transition is pushing all of our unrendered baggage into the light. Integration is no longer just a path to self-improvement; it is a structural necessity to prevent major destabilization.

Radical Honesty: The Price of Sovereignty

If the Shadow is essentially "everything we are afraid to look at," then the act of Sovereignty begins with a single, terrifying act: Honesty.

Most people are living in a state of social camouflage. They fit in with the crowd and they suppress their true spirit because they are afraid of being seen. This is the ultimate Cork behavior. By trying to fit in, you are intentionally dampening your own signal. You are telling the Universe: "I am not a unique wave; I am just more of the water."

Stepping into the Shadow means Stepping into Fear. Integration is not a moral process; it is a biological calibration. When you accept every part of yourself, you are taking that negative pole and hooking it back up to the positive pole. You are creating a closed circuit.

The Courage of the Wave

Make no mistake: becoming a Soliton takes courage.

It is the courage to be "whole" rather than "perfect." In a world obsessed with fixing the parts, standing as a Unified Whole is a radical act of rebellion. Courage is the non-linear fuel of the Soul. Without it, you cannot generate enough internal intensity to counter the dispersion of the world. You cannot lock the wave without the heat of your own bravery.

The Free Spirit and the Invincible Wave

A Free Spirit is often misunderstood as someone who is flighty. In reality, a Free Spirit is the most structured entity in existence because their structure is Internal.

- A Linear Wave is well-behaved. It follows the crowd and disperses when it hits an obstacle.

- A Non-Linear Soliton is rebellious by nature. It holds its own shape. It passes through obstacles without losing integrity.

This is the Sovereign state. It is the realization that your fear is not a sign to stop; it is the dispersion term in your equation. When you step into the fear, you are providing the necessary resistance for your signal to lock. You are using your darkness to sharpen your light.

Loving the Whole: The Handshake of Integrity

Integration is the ultimate act of self-love. The Fixer mentality says: "I will love myself when I have fixed my anxiety." The Sovereign mentality says: "I love the anxiety because it is the 'Negative Pole' that allows my energy to flow."

When you love every part of yourself without embarrassment, you stop being a leaky system. Your energy is no longer spent on the Fixer's endless war of suppression. Instead, all that power—the power of your Shadow and the power of your Light—is channeled into a singular transmission. You become immutable. You move through the Ocean of the Aether with a different kind of authority.

Chapter 6: Coherent Polarity

The Inner Alchemy of the Solar and Lunar

The Subtle Trap of Separation

There is a common and highly understandable phase in the journey of awakening. As we begin to break away from the rigid, linear structures of the mainstream world, we naturally gravitate toward spiritual communities. These spaces offer profound healing, shared insights, and a sense of belonging. Yet, if we observe the mechanics of these groups closely, a subtle dynamic often takes root. An unintentional boundary forms—a quiet polarity between those who are "spiritual" and those who are not; between those "doing the work" and those who are asleep.

Without realizing it, this "us versus them" mentality acts as a gentle continuation of the very religious paradigms of separation that many in these communities quietly critique.

This habit of separation doesn't just happen outwardly between groups of people; it bleeds into how we view our own internal architecture. Within this movement, it is very common to hear terms like "divine masculine" and "divine feminine." These are beautiful, ancient archetypes that hold deep truth. However, the trap emerges when the culture of separation turns these internal frequencies into externalized objects.

An individual might do a significant amount of deep inner work, only to immediately look to the outside world to find their matching "divine" counterpart. They are inadvertently taught that they must find this missing piece in another person to complete their energetic circuit.

While the dynamics of a polarized relationship are incredibly powerful, externalizing these forces is a trap for anyone truly seeking personal sovereignty. It creates dependency. It teaches individuals to remain incomplete circuits, perpetually searching for an external partner to validate their wholeness.

In the ancient Hermetic traditions, and later in Carl Jung's deep work on alchemy, the ultimate goal was never to find your elemental opposite in another person. The goal was the *Mysterium Coniunctionis*—the Alchemical Marriage of the solar king and the lunar queen inside the exact same crucible. To achieve true sovereignty, you must learn to generate your own power by balancing these two fundamental forces within a single vessel.

Scale Invariance and the Elemental Archetypes

To understand these forces, we strip away cultural baggage and look directly at the hard physics of nature. Through the universal law of Scale Invariance —the mechanical proof of As Above, So Below—the exact same polarity we see in the cosmos maps directly to elemental metals and human psychology.

The Sun (the stable, radiant, unyielding center) corresponds to the noble metal of gold and the masculine archetype. The Moon (the reflective, dynamic, moving body) corresponds to the highly reactive metal of silver and the feminine archetype.

The Sovereign Masculine: The Solar Anchor

In metallurgy, gold is a "noble" metal. It is incredibly dense, virtually indestructible, and chemically inert —meaning it does not react to oxygen, water, or most acids. It does not rust, tarnish, or decay. It holds its exact geometric state regardless of the harshness of its environment.

Translated to consciousness, this is the Solar Anchor.

- Non-Reactive Sovereignty: Your inner solar structure is your structural integrity. It is the part of your consciousness that observes without reacting and maintains the baseline. When the macro environment gets frantic, this immovable anchor simply says, "I hold my state."

- The Crucible: In physical alchemy, you cannot melt and transform highly volatile elements without a heavy, heat-resistant crucible. Your inner masculine is that vessel. It creates a rigid, safe psychological perimeter where intense experiences and emotions can burn without destroying the host.

The Conductive Feminine: The Lunar Current

In physics, silver is the ultimate transmitter. It possesses the highest electrical conductivity, the highest thermal conductivity, and the highest light reflectance of any element. Just as the Moon dynamically interacts with the Earth to pull the tides, silver is highly reactive, constantly engaging with its environment and acting as a flawless mirror.

Translated to consciousness, this is the Lunar Current.

- Absolute Conductivity: Your inner lunar energy is your interface with the living universe. It is the high-voltage current that flows through the structure you built. It is your raw sensitivity—the exact part of you that picks up geopolitical tension, the unspoken energy in a room, or the broader patterns in the Aether.

- The Fluid Adapter: Where the solar structure is rigid, the lunar current is malleable. It gives you the capacity to shift strategies, accept new truths, process grief, and creatively navigate a changing world. It doesn't block emotion; it flows it perfectly.

The Thermodynamic Loop

When you split these roles, the system fails. If you have pure lunar conductivity without the solar structure, the high-voltage energy has nowhere to ground itself, leading to emotional burnout or chaotic dissipation. Conversely, if you have pure solar structure without the lunar current, you have a perfectly secure, unmovable vault—but it is completely dark and lifeless inside.

When you integrate both within your own vessel, you create a frictionless thermodynamic loop. Your inner anchor builds the secure perimeter, granting your inner conductor the absolute safety to feel, explore, and transmit energy without the fear of being overwhelmed. Simultaneously, your conductive energy infuses your rigid structure with life, preventing your anchor from becoming a dead, isolated vault.

The Soliton as the Unified Vessel

This brings us directly back to the physics of the Soliton. A Soliton wave maintains its impossible stability by perfectly balancing two opposing physical forces:

1. Dispersion (The Solar Constraint): The structural property of the medium that forces a wave to naturally spread out, flatten, and dissipate into nothingness. It acts as a dampening, regulatory constraint.

2. Non-Linearity (The Lunar Flow): The dynamic, high-velocity force that causes a wave to

want to steepen, compress, and explosively sharpen. It is pure localized power, movement, and conductive flow.

If you have non-linearity without dispersion, the wave instantly sharpens into a chaotic breaking point and crashes. If you have dispersion without non-linearity, the wave simply flattens out and dies.

A Soliton only emerges when the dynamic sharpening of the wave precisely matches the structural dampening of the medium. The energy and the structure lock into a perfect, self-sustaining embrace.

When an individual achieves this coherent polarity, they aren't trying to be an idealized, static version of a human being. They have locked their inner sun and moon into a dynamic equilibrium. When the macro system around us experiences extreme friction—when the credit lines crack, the geopolitics fracture, and the collective mind panics—the un-integrated individual is instantly torn apart by the waves of external energy. But the person who has locked their dual archetypes into a Soliton wave simply passes through the chaos entirely intact. They are no longer an incomplete circuit looking for an external stabilizer. They load their own reality on demand.

Chapter 7: The Interface

The Observer's Gaze and the Loading of Reality

The Quantum Handshake

In the early 20th century, physics hit a wall that changed everything: the Double-Slit Experiment. It revealed that electrons—the building blocks of what we consider solid matter—behave like waves of potential until they are observed. The moment a conscious "witness" is introduced, the wave collapses into a localized particle.

This is the Quantum Handshake. It suggests that the universe is not a finished product sitting on a shelf; it is a state of infinite, fluid potential that takes a specific, tangible form only when a conscious signal interacts with it.

The Video Game of Reality

To understand how this functions in our daily lives, we can look to the architecture of high-end digital simulations. In a modern video game, the engine doesn't waste energy rendering the entire world at once; it only "loads" the objects that are within the player's field of vision.

The universe operates on the same principle. Think of it like the Field of View in a high-end simulation. The Aether spins up into matter only where the Observer's gaze is fixed. Behind your head, the world remains as code (the Wave); in front of your eyes, it collapses into a 3D environment (the Particle).

The Flicker of the Frames

Think of this "Loading on Demand" as a series of high-speed frames in a film. Just as a movie is actually a sequence of individual still photos flickering past the lens to create the illusion of smooth motion, your reality is a sequence of "Now" moments loading at a near-infinite frequency.

You aren't moving through a solid, static world; you are witnessing a rapid-fire rendering of frames. When your internal noise floor is high, these frames blur together into a thick, unchangeable fate. But as you integrate your shadow and find presence, you start to sense the space between the frames. You realize that because the render is fresh in every single moment, you are never truly stuck in a solid past.

The Physical Now: The Render Point

Many talk about the "now moment" as a psychological concept—a way to find mental peace, famously explored in works like Eckhart Tolle's The Power of Now. While this is a helpful framing for the mind, what we are discussing is the Physical Now.

The Interface is the realization that there is nothing but the Now because the Now is the Render Point. It is the nanosecond where the "software" of the Greater Mind becomes the "hardware" of the physical world. The red rocks or the ground beneath your feet are a continuous stream of information being rendered in real-time, frame by frame.

Calibration: The Balance of Silence and Viscosity

To reach this level of awareness, we need both Silence and Viscosity.

Silence: Alone time and contemplation are essential for improving self-awareness and lowering the background noise of life.

Viscosity: Interaction with the world—relationships and challenges—provides the necessary friction or viscosity required for transformation.

A Personal View: Witnessing the Engine

When the internal noise floor drops low enough, the Interface reveals its grain. In states of deep, integrated silence, I have witnessed the engine running behind the surface of the physical world.

I perceive this as a field of extremely fast-spinning, thin standing waves. To understand this, imagine a piece of fine, taut silk stretched over a massive hurricane. While the world generally only perceives the smooth, finished surface—the solid physical reality—I am witnessing the render point. I am seeing the exact moment where the invisible power of the spinning storm pulls at the fabric of existence, revealing an intricate, spinning geometry beneath the silk of our reality.

These are interacting spiral vortices rippling through the fabric of the Now. I am able to perceive the wave pattern behind the silk—the high-frequency vibration that organizes the medium just before it collapses into the three-dimensional forms we touch. What we call "solid" is simply the state where the hurricane holds the silk so tightly that the underlying wave becomes invisible to the ordinary eye. But in deep states of presence, I am able to see the engine spinning up the frames, witnessing the transition from pure energy into the persistent illusion of matter. This spinning field—this standing wave holding its shape beneath the render is a Soliton waveform. The entirety of this book exists because the Aether used those states of deep presence to reflect that living architecture back to me.

The image below is a visual approximation of what stillness reveals; the Soliton. Paper cannot capture the depth complexity.

The Pre-Render Soliton — Still Point

This is the visual proof of the "Handshake." It is witnessing the exact moment where internal frequency meets the field and "spins" it into the appearance of a solid object.

The Geometry of the Spin: A Universal Constant

Why a spiral? From the macroscopic to the microscopic, reality is defined by Spin.

The Macro: Galaxies are massive, spinning spiral vortices.

The Planetary: The Earth spins on its axis and follows a helical (spiral) path through the galaxy.

The Micro: Protons and electrons possess "spin"—a fundamental property of their interaction with the field.

The spiral is the most efficient way for the fluid Aether to organize itself into a stable form. It is the geometry of the Vortex—the point where the Infinite is squeezed into the Finite.

Nuanced Loading: The Higher Journey

"Loading on Demand" is not an egoic manifestation tool; it is a subtle dialogue with your Higher Self. This higher aspect of your consciousness loads exactly what is best for your growth. Sometimes it loads ease, and other times it loads difficulty—the viscosity needed to sharpen your signal. The Sovereign Being doesn't try to force the render; they operate with a higher awareness of what is loading and why.

Chapter 8: The Geometry of Presence

The Hub, the Rim, and the Still Point

The Map of the Spin

In the previous chapter, we witnessed the engine of reality—that extremely fast-spinning, thin standing wave pattern that renders the world frame by frame. While seeing the spiral is a profound experience, the question remains: How do we navigate it? To live as a sovereign being, you need a map. That map is the Wheel.

Whether it is the "wheel of the year," the "wheel of karma," or the rotation of a galaxy, the geometry remains scale-invariant. By understanding the relationship between the center and the edge, you can choose where you stand in the Now moment.

The Rim: The World of High Velocity

The Rim of the wheel represents the external world. It is the realm of events, deadlines, birth, death, and social drama. This is where the "noise" is at its maximum.

In physics, the linear velocity (v) of a point on a rotating disk is determined by its distance from the center (r) and its angular velocity (ω):

$$v = \omega r$$

On the Rim, r is at its maximum. This means the linear velocity is at its highest. When you identify with the rim—when your happiness depends on external circumstances—you are moving at a breakneck speed. Every bump in the Ocean of the Aether is felt with maximum force. This is the habitat of the Cork; you are being tossed around by a rotation you cannot control.

The Hub: The Still Point

At the exact center of the wheel lies the Hub. If we look at our equation again, at the center, $r = 0$. Therefore:

$$v = \omega(0) = 0$$

The Hub is the still point. It is the seat of the Witness. While the rest of the wheel is spinning at incredible speeds, the center is fundamentally at rest. This is the goal of shadow integration and presence: to shift your identification from the high-velocity chaos of the Rim to the zero-velocity silence of the Hub.

Standing at the Hub does not stop the world from spinning. The bills still need to be paid, the body still ages, and the seasons still change. However, from the Hub, you are no longer being tossed around. You are the Conductor, observing the symphony without being overwhelmed by the volume.

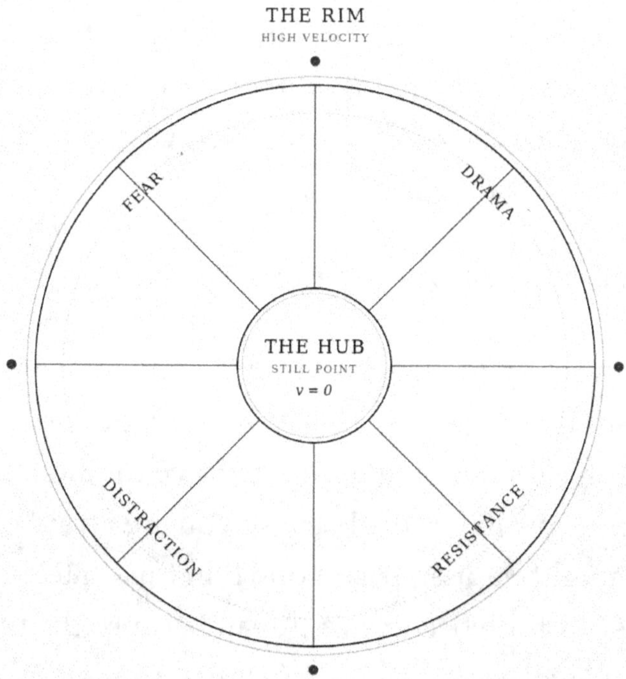

The Spokes: The Tension of Integrity

The Spokes are the connectors. They represent your signal—the actual transmission of your intent from the Hub out to the render on the Rim. In a bicycle wheel, the rim only stays round and the hub only stays centered if the spokes are under equal tension. This is where the practice of mastering nuance begins.

In our current paradigm, we are taught to live in extremes. We are told to either fight for what we want (excessive tension) or to give up and let life happen to us (lack of tension). Neither of these states creates a Sovereign Soliton.

I. The Danger of Over-Tension (The Force)

When you try to manifest from the ego, you are over-tightening the spokes. This creates a rigid, brittle structure. You are attempting to force the Aether to render a specific frame because you are afraid of what might happen if you don't. In this state, the viscosity of life becomes a threat. Because the wheel has no give, every bump in the road sends a shockwave through the hub. Eventually, the spoke—or the person—snaps.

II. The Danger of Under-Tension (The Drift)

Conversely, if the spokes are too loose, you become the Cork once again. You have no signal to organize the field. You are drifting, and the wheel of your life becomes out of true. It wobbles, it drags, and it eventually collapses because there is no coherent frequency holding the structure together.

III. The Nuance of Surrender

Mastering the spokes is about finding the optimal load. This is the Nuanced Manifestation we alluded to in the previous chapter. It is the understanding that you are not forcing the render; you are optimizing it through a state of Active Surrender.

Active Surrender is not passive. It is the highly focused act of maintaining your frequency (your Hub) while allowing the Greater Mind to provide the specific details.

Setting the theme of the frame is not an intellectual request or a mental wish list; it is a resonant state of knowing. If you use the ego to visualize abundance, you are often just broadcasting the tension of your own desire. In the physics of the soul, you do not get what you want; you render what you are.

To "set" a theme, the internal oscillator—the heart-brain handshake—must generate the coherent feeling of the quality as an existing fact. You are not asking the universe to change the render; you are changing the signal so that a new render becomes the only logical outcome. It is the move from seeking a destination to being the frequency of the destination. You provide the resonant theme, and you surrender the how, trusting that a coherent signal will always organize the Aether into a coherent reality.

The universe responds Quantumly, not linearly. This means the first thing to load in your render is often a challenge that feels like the exact opposite of your intent.

This isn't a glitch; it is a frequency test. It is a non-linear setup designed to see if you can hold your Hub and your frequency even when the render gets difficult. In the Quantum world, stability isn't given; it is earned through resonance. If you can maintain your signal through this initial friction, the system phase locks, and the new reality finally stabilizes around you. You are learning to navigate the non-linear landscape of the field.

The Spiral of Time: The Wheel in Motion

To conclude the geometry, we must realize that the wheel is never just sitting still. As it moves through the ocean of the Aether, it creates a spiral.

If you look at a wheel from the side, it is a circle. But if you track a single point on that wheel as it moves forward through time, that point traces a helix. This is why I perceive the interface as a spiral vortex. The thin standing wave is the Wheel of your Presence spinning at near-infinite velocity as it travels through the field of time.

When you stand at the Hub, you are at the center of that spiral. You are the Still Point around which the entire universe is rendering its dance. You aren't chasing the future or mourning the past; you are the oscillator at the heart of the spin, witnessing the frames as they load in perfect, nuanced sequence.

Chapter 9: The Master Code

The Architecture of the 98%

The Wisdom of the Wind

Imagine a structure rising one kilometer into the sky—the Jeddah Tower in Saudi Arabia. At such an unprecedented height, the building's greatest adversary isn't gravity; it's the wind. For a standard building, a strong wind is a destructive force that causes dangerous swaying. Most skyscrapers try to survive this by fighting back—using massive weights called dampers to counteract the movement.

But the Jeddah Tower uses a more sovereign strategy. It is designed with a tri-petal, tapered footprint. This specific geometry doesn't fight the wind; it uses it. As the wind hits the tower, the "wings" break up the air currents, preventing them from forming the destructive oscillations that would cause the building to sway. The wind's own pressure actually helps stabilize the structure.

The Soliton Strategy: From Fixing to Integrating

When life's problems show up for a Soliton, they don't pretend the struggle doesn't exist. They still feel the pain, the fear, and the pressure—just as the tower feels the wind. The difference is in the geometry of perspective.

A Cork sees a problem and immediately tries to fix it. They want the wind to stop blowing. They exhaust themselves trying to change the external world so they can feel safe again.

A Soliton, however, does not try to fix the wind. They sit with the pressure and ask:

- What is this showing me about my own structure?
- How can I use this friction to sharpen my signal?
- What part of my "Shadow" is this wind catching?

By looking at the challenge from a higher vantage point, you redirect the energy. You aren't just surviving the problem; you are using the viscosity of the situation to lock your presence even deeper into the Now. When you think this way, you are moving off the chaotic rim and beginning to live on the spokes.

The 98% and the Aether

To understand why this structural flip works, we have to revisit the 2% and the 98%. As humans, our senses only perceive about 2% of reality—the material world of "things." But the other 98% isn't empty space; it is a high-pressure, super-fluid ocean of potential, known as the Aether.

In physics, the Navier-Stokes equations describe how fluids move. For over a century, there has been a million-dollar prize for anyone who can solve these equations. But perhaps they aren't meant to be solved by the mind; they are meant to be understood by the Witness.

You don't "solve" the ocean to stay afloat; you understand its quality and resonance. Ibrahim Karim, the founder of Biogeometry, describes this Aether as the "Primordial Water." This isn't just a metaphor. It is the original medium of creation. You learn to navigate the 98% (the quality) so that the 2% (the physical render) of your life stays stable.

The Great Year: The 26,000-Year Reset

We are currently living through one of the most significant moments in human history. This is because we are completing a full circuit of what is known as the Great Year—a cycle of approximately 26,000 years caused by the slow wobble of the Earth's axis.

As the Earth moves through this cycle, we pass through different "Ages," each lasting about 2,150 years. Each age acts as a different theme for the human experience:

- The Age of Leo (approx. 10,000 BCE): The era of the Sphinx and the solar consciousness.
- The Age of Aries: The era of the individual, the warrior, and the law.
- The Age of Pisces: The era we are currently leaving. It was symbolized by the fish—submerged in the water, subject to the hidden currents of the subconscious, and often defined by sacrifice and religious dogma.

The Age of Aquarius: The Return of the Bearer

We have now reached the end of the entire 26,000-year spiral. We are moving into the Age of Aquarius, the Water Bearer.

In the Piscean age, we were the fish in the water. We were immersed in the medium but unaware of it, subject to its currents. But the Water Bearer—depicted as a woman pouring from a vessel—is the one who has mastered the medium. She is not in the water; she is bearing it.

The Age of Aquarius is the era of the 98%. It is the shift from quantity to quality. The water she pours is the Aetheric Fluid—the information and intent that renders our physical world. Because we have completed a full cycle, we are finally being given the Master Code to understand the medium we've been swimming in for millennia. We are moving from a state of being pushed by life to bearing the light of our own intent.

The Sovereign Flip: Who Chose the Wind?

Once you have mastered the art of using the wind to stabilize your tower, a deeper question arises: Are you simply learning to use the wind, or are you choosing the wind to refine your structure?

In the linear world, things happen to us. But in the 98%, you are coupled to the field. Your internal spin and the external flow are one system.

Free Will is the power of resonance. When your internal frequency shifts, the render of the Aether shifts with you. The difficulty you are facing isn't a random accident; it is the specific viscosity you have requested to help lock your geometry. You are the Architect, the Engineer, and the Tower all at once.

Chapter 10: The Fractal Pyramid

Scale Invariance and the Earth-Vessel

The Precision of the Signal

In my thirty years working with high-precision scientific instrumentation, I have learned that the integrity of any result depends entirely on the quality of the tuning. It isn't enough for a signal to simply appear; it must be perfectly aligned with the instrument. Mastery of any instrument requires a consistent focus on tuning, calibration, and maximizing the signal-to-noise ratio.

Tuning is the deliberate refinement of operating conditions so the instrument resonates with the physics it is measuring. When we look at the Great Pyramid of Giza through this lens, we see a masterclass in resonant engineering designed with calibration that rivals our most advanced modern sensors.

The Tuned Granite of the King's Chamber

As the engineer Christopher Dunn has meticulously documented, the architecture of the Great Pyramid shows signs of intentional, high-precision tuning. Above the King's Chamber, in the series of five "Relieving Chambers," Dunn identified massive granite blocks where material appears to have been intentionally shaved away. From a technical perspective, this suggests these blocks were being "tuned" to a specific acoustic frequency. The builders understood that for this stone vessel to function, its internal hardware had to be calibrated to a singular, coherent frequency that matched the peak of the planet's own resonance. They were cleaning the internal noise of the structure to maximize its sensitivity to the Aetheric field.

The Piezoelectric Core: Quartz as a Transducer

To understand why the ancients used granite for the inner chambers, we must look at the mineralogy of the stone itself. The granite used in the King's Chamber was transported from Aswan, over 500 miles away, specifically because of its high quartz content. In modern instrumentation, we use quartz for its piezoelectric properties—the ability to convert mechanical pressure into electrical energy. The King's

Chamber is surrounded by millions of tons of limestone. This immense overburden creates a constant mechanical stress on the granite core. In effect, the King's Chamber is a massive, solid-state transducer. It wasn't designed to be a static room; it was designed to be a vibratory engine, turning the tectonic and atmospheric "noise" of the Earth into a coherent, high-frequency signal.

The Eight-Sided Geometry: The Concave Secret

One of the most overlooked details of the Great Pyramid is that it is not actually four-sided; it is eight-sided. Each of the four faces is slightly concave, a feature only visible from the air during the equinoxes. This slight indentation is not a structural flaw or the result of erosion; it is a calculated geometric necessity.

In physics, a concave surface acts as a parabolic reflector. By curving the faces inward, the builders created a structure that could focus incoming energy —whether solar, atmospheric, or Aetheric—directly into the center of the monument. This concave geometry gathers the field, further increasing the signal-to-noise ratio at the core where the observer sits. It is the architectural equivalent of a satellite dish, tuned to the frequency of the Earth itself.

The 43,200 Ratio: A Model of the World

The Great Pyramid is a 1:43,200 scale model of the Northern Hemisphere. If you take the height of the pyramid and multiply it by 43,200, you arrive at the polar radius of the Earth. If you take the perimeter of the base and multiply it by that same number, you get the equatorial circumference of the planet. By building at this specific ratio, the ancients ensured the pyramid was "phase locked" with the Earth. It was a physical transducer designed to catch the low-frequency hum of the planet and step it down into a frequency that the human nervous system could perceive.

Reading the Code: The Consciousness Interface

If we accept that the Great Pyramid is a perfect 1:43,200 scale model of the Earth, we are forced to confront a staggering question: How did the builders know the exact dimensions of the planet—the polar radius and the equatorial circumference—long before the advent of satellite geodesy?

From the perspective of a strictly linear materialist history, this is an impossibility. But if we view reality through the lens of the simulation, the answer becomes clear. In a "Loading on Demand" universe, the physical planet is the hardware, but the mathematical blueprints for that hardware exist in the 98% Aetheric Field. The builders did not need to physically measure the Earth with sticks and shadows; they measured it through Consciousness.

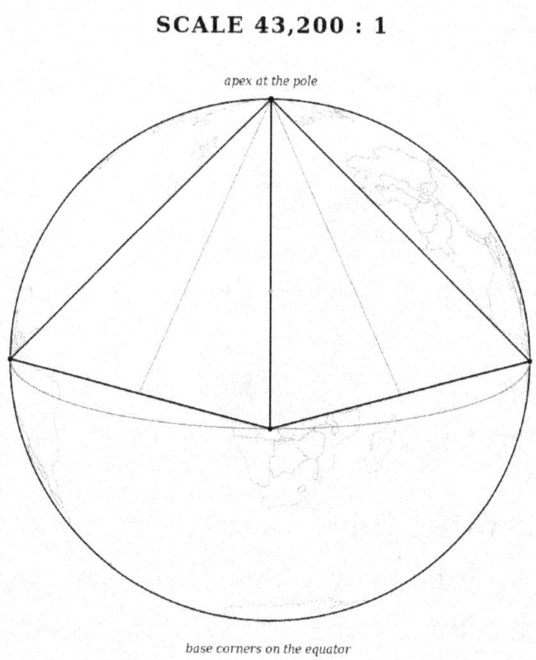

The Fractal Pyramid

The Orion Handshake: The Galactic Directory

This ability to "read the code" extended far beyond the atmosphere of the Earth. The layout of the three Giza pyramids serves as a terrestrial mirror for the three belt stars of the Orion constellation as they appeared thousands of years ago.

This is the ultimate "As Above, So Below" handshake. By aligning the base of the monument with the stars and its dimensions with the Earth, the ancients created a bridge between the cosmic and the planetary. They understood that the Aether is a single, continuous fluid. To influence the render on Earth, you must be in resonance with the field of the stars. They were essentially mapping the Galactic Metadata onto the local hardware of the Giza plateau.

The Master Code and the Great Year

As we discussed earlier, we are currently finishing a 26,000-year circuit of the Great Year. Just as our year has seasons, the Great Year has "seasons of consciousness." In the era when the Giza complex was rendered, the Aetheric weather was different. The noise floor of the planet was lower, and the viscosity of the medium was such that the human mind could interface directly with the source code of the reality-engine. These architects were system administrators who could download planetary and stellar dimensions with the same ease that a modern programmer views the properties of an object in a digital simulation.

Conclusion: The Macro-Vessel

The Great Pyramid serves as the macrocosmic version of your own biology. It is a Vessel (the stone) and a Signal (the tuning) moving through an Ocean (the Aether). By studying this ancient machine, we realize that the laws of sovereignty aren't just spiritual concepts—they are the literal building blocks of the universe. To be sovereign is to be as stable, as tuned, and as perfectly scaled as the pyramid itself. It is a stone monument to the fact that the world is not a solid, accidental endeavor—it is a purposeful, mathematical expression of Mind.

Chapter 11: Divine Intelligence and the Fractal Offset

The Wisdom of the Hoya and the Deep Roots

The Invisible Architecture

Sometimes, the universe does not communicate in grand, sweeping cosmic events; it speaks through the quiet, geometric persistence of nature. I have a Hoya plant in my home. For five years, it appeared to be doing very little—just a static system producing green leaves but absolutely no flowers.

But nature is never truly static. Beneath the surface, the plant was spending years meticulously building and expanding its root system. It was establishing the deep, structural integrity required before it could support the immense energy of blossoming outwards.

Then, at the beginning of May 2026, something profound shifted within its biological render. Having secured its roots, it suddenly produced small, perfectly pentagonal buds. They appeared thick, highly structured, and almost alien, clustered together at the end of a delicate stem to form a perfect, downwards-pointing umbrella or dome shape.

When I counted these newly rendered structures, there were exactly 19 of them. Nature does not make mistakes. The plant didn't randomly produce 19 buds. It was communicating a specific structural necessity—a mathematical frequency—right into my living space.

The 19 Pentagons of the Hoya

The Geometry of the Blueprint

To understand this communication, we must look at the geometry of the plant's render. The buds first appeared as pentagons. A pentagon is entirely defined by the Golden Ratio (Φ), where the ratio of a diagonal to a side is exactly **1.618**. When these buds opened into five-pointed stars, every intersection of their lines divided the segments into that exact same Golden Ratio. The downward-facing umbrella cluster was the necessary arrangement allowing these individual fractal pentagons to work together as a single, physical fractal antenna.

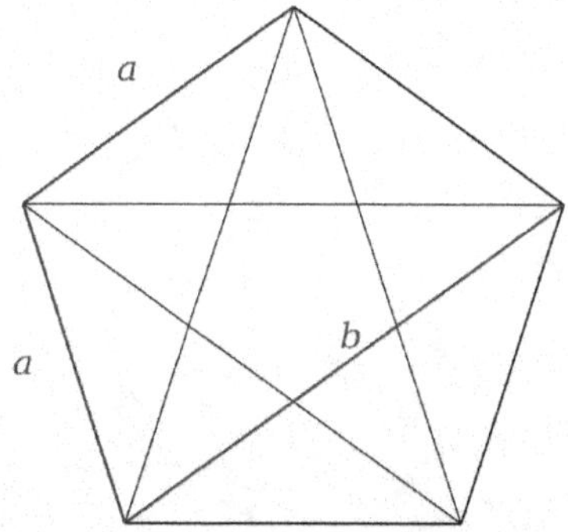

Ratio $\frac{b}{a} = 1.618...\ (\Phi)$

The Golden Ratio in the Pentagram

But the true masterstroke of Divine Intelligence was the quantity. There were 19 buds.

In esoteric geometry, the complete blueprint of the unified field—the Flower of Life—is built from exactly 19 intersecting circles. The Hoya plant was physically demonstrating the complete matrix of creation.

The Flower of Life

nineteen interlocking circles

The Engine of Evolution: 19^2 and the +1

The significance deepens when we look at the mathematics of that number. If you square 19, the result is: $19^2 = 361$.

In the physics of our reality:

- **360** represents the completed cycle, the full circle, the static loop. It is closed. In fluid dynamics, a closed loop just spins in place forever.
- **361** breaks the static loop. It is **360 + 1**.

If you add that 1—a slight offset or a push into the next dimension—that closed circle becomes a spiral. That +1 is the engine of evolution. It takes the complete blueprint of reality (the 19 circles) and spirals it upward into the next octave of growth.

For five years, my plant was locked in a 360-degree holding pattern, building its foundation. When it was finally ready to evolve, it utilized the Divine Intelligence of the +1 offset to break the loop, pushing a static blueprint into a living, spiraling geometry.

The Microscopic +1: The DNA Offset

To truly grasp the power of this offset, we have to look at how nature builds it into our own physical Vessels. To bridge the gap between the infinite, theoretical Φ and finite, physical matter, nature uses the Fibonacci sequence (1, 1, 2, 3, 5, 8, 13, 21, 34...). Every number in this sequence is a whole integer, making it perfect for building physical structures. When you divide adjacent Fibonacci numbers, they approximate Φ, oscillating back and forth but never actually touching the exact number.

We see this ultimate +1 engine in the very architecture of human DNA. A single full cycle of a DNA double helix measures 34 angstroms long by 21 angstroms wide. If we divide these two Fibonacci numbers, we get:

$$\frac{34}{21} \approx 1.619$$

Notice that it is not the perfect 1.618 of the Golden Ratio. If DNA were built on the exact, perfect mathematical ratio, the structure would be perfectly balanced. It would close a perfect loop, lock into place, and become totally inert.

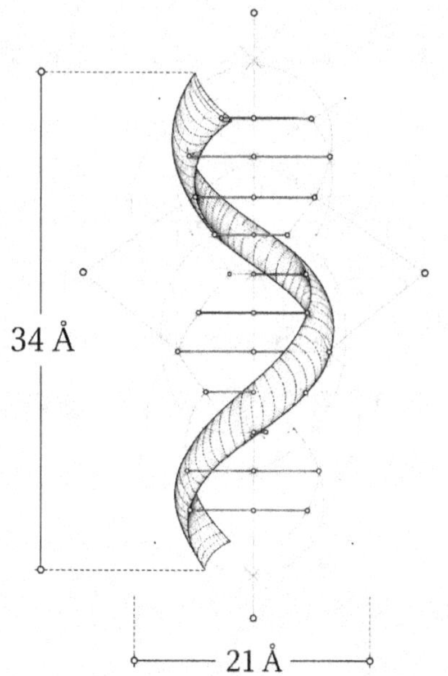

$34 / 21 \approx 1.619$

The 1.619 Fibonacci Offset — The Engine of Life

The fact that DNA operates at 1.619 creates a microscopic geometric offset. That tiny differential creates a persistent structural tension, forcing the molecular chain to twist into a 3D double helix rather than remaining a flat 2D circle. It is the geometric driver of life itself.

The Divine Tension

Whether it is a Hoya plant spending years building its root system before using the +1 offset to blossom as a structured dome, or the **1.619** ratio forcing our

DNA to twist into a living spiral, the lesson is clear. The universe is not composed of perfectly closed circles. It requires this geometric offset—this Divine Tension—to continually spiral energy forward. When we understand this, we realize that the challenges, delays, and frictions in our own lives are not mistakes; they are the +1 engine forcing our consciousness to evolve.

Chapter 12: The Biological Render

The Biological Render and the Master of Health

The Inner Ocean: The 80% Medium

We often think of our bodies as solid, permanent objects. But if we look closer, we realize that we are essentially walking, talking columns of water. Our bodies are roughly 70% to 80% water—which means we are more medium than object.

This connects back to our "Primordial Water" in the previous chapter. Your body isn't just in the Aether; your body is made of it, compressed into a biological form. Because water is highly conductive and responsive to frequency, your body is the perfect vessel to receive the render. Every cell is bathed in this inner ocean, waiting for the signal to tell it what to become.

The High-Frequency Refresh Rate

To understand health, we have to look at the refresh rate of reality. Think of a high-definition television screen. The image of a person moving across the screen looks solid and continuous, but it is actually a series of individual frames being refreshed at 60 or 120 times per second. If the signal is clear, the image is crisp. If there is interference, the image glitches.

Your body works the same way. It is a High-Frequency Render. Every second, millions of your cells are replaced, and every one of those replacements is a New Frame loading from the Aether. You are not a static object that gets old; you are a persistent pattern of information being refreshed in every Now moment.

The Heart: The Master Oscillator

In the physical vessel, the primary oscillator is your Heart. The heart is far more than a mechanical pump; it is the most powerful generator of electromagnetic energy in the body. It creates a massive, toroidal (donut-shaped) field that extends several feet away from your skin.

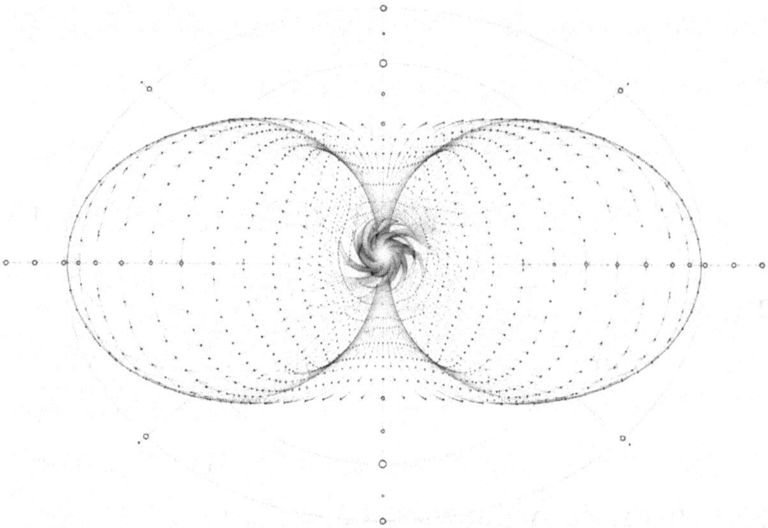

The Toroidal Field of the Heart

The heart's job is to "imprint" your current frequency onto the 80% water within you. When your heart is in coherence, it sends a clean, powerful signal to every cell. This signal acts as the Master Blueprint for the next frame of your biological render.

Dynamic Equilibrium: Staying in the "Lock"

Health, from a Soliton perspective, is a state of Dynamic Equilibrium. In physics, this describes a system where opposing forces are in perfect balance —like a skyscraper standing still despite the wind.

You are never "done" with health. It is a continuous act of re-balancing. The world provides the Viscosity (the stress, the toxins, the noise), and your consciousness provides the Non-linearity (the focus and intent). Mastery is the ability to maintain this equilibrium frame after frame.

Beyond "Positive Thinking": The Structural Work

There is a common misunderstanding that you can change your health simply by "thinking positively." While intention is a key variable, relying on surface-level positive thoughts is like trying to stabilize the Jeddah Tower by painting it a pretty color! You can paint the building bright yellow, but if the underlying geometry isn't integrated to handle the wind, the structure will still sway.

True health is a result of Structural Integration. It is a mastery that is cultivated over years of contemplation and shadow work.

- Shadow Integration: This involves sitting with the "wind"—the trauma, the fear, and the suppressed parts of yourself—and integrating those forces into your geometry so they no longer create “drag” in your signal.
- The Long Game: While everyone has to start somewhere, you don't overwrite a lifetime of "noise" with a single weekend of Zen. It takes

years of consistent effort to clean the lens of your projector so that the next frame loads with clarity.

The Sovereign Architect

As you navigate your own path—whether through the landscape of your daily life or the quiet moments of your own mind—realize that every Now moment is a choice. You aren't just moving a body; you are a Sovereign Oscillator managing a biological render. You are the Architect, and the quality of the next frame depends on the depth of your integration.

Chapter 13: The Medium

The Ultimate Phase Change

The Cosmic Joke: Tuning into the Medium

It is one of the universe's most enduring cosmic jokes that we call people who communicate with the "other side" Mediums. Usually, we assume this means they are in the middle of two worlds. But in the physics of Sovereignty, the word is much more literal.

If you are a Soliton—a stable wave of information—you are always moving through a Medium. For the duration of your physical life, that medium is the 80% water of your biological vessel and the high-pressure fluid of the Aether. When we die, we aren't going to a different place; we are simply changing our relationship to the Medium.

A "Medium" is just someone who has tuned their antenna to the frequency of the "Steam" instead of just the "Ice." They haven't left the building; they've just moved from the noisy lobby to the high-speed, invisible data-stream of the 98%.

The Physics of Graduation: Ice to Steam

To make sense of this "completion," we return to our inner ocean. Throughout this book, we've worked to move from being Ice (the rigid, fearful "cork" state) to being Water (the fluid, integrated Soliton).

But what happens when the frequency of experience reaches its peak? In physics, this is a Phase Change.

- Ice is limited, heavy, and separate.
- Water is flowing, connected, and powerful.
- Steam is expansive and invisible to the 2% senses, yet it moves through everything.

Death is the moment the Soliton signal completes its journey through the viscosity of the physical world and transitions into the steam phase. You don't lose your "information"—your memories, your integrated shadow, and your refined signal. You simply lose the drag of the 2% render. You become a high-frequency broadcast moving through the 98% Aether without the need for a physical antenna.

The Fear of Evaporation: The Ego's Last Stand

If this graduation is so triumphant, why are we so terrified of it? The answer lies in the Ego.

The Ego is the part of us that identifies exclusively with the Ice. It likes the solid walls. it likes the feeling of being a "thing" that is separate from other things. To the Ego, the prospect of becoming steam doesn't look like freedom; it looks like evaporation. It fears that if it loses its hard edges and its 2% meat suit, it will simply vanish into the void.

But this is a misunderstanding of the physics. When water evaporates, it doesn't cease to exist; it actually becomes Omnipresent. It is no longer stuck in one glass or one puddle; it is now part of the atmosphere, capable of moving through mountains and riding the winds of the entire planet.

The work of years of contemplation is essentially the process of convincing the Ego that it is safe to melt. By integrating your shadow, you are teaching your consciousness that you can lose your Ice identity without losing your signal. You are practicing the art of being Steam while you are still in the Water.

The Conservation of the Signal

One of the most iron-clad laws of the universe is the Conservation of Information. Mainstream science knows that energy cannot be created or destroyed, but a Soliton knows that Identity (the Master Code) is also conserved.

Think of your life like a high-end video game. When the console (your body) finally gives out, the character you've spent years building doesn't vanish. The "save file" is held in the Cloud—the 98% Aetheric memory.

The completion is the realization that you aren't the character on the screen; you are the Information that persists even when the screen goes dark. You are the 100%—the signal and the medium as one.

The High-Frequency Wake

If we treat death as a Phase Change rather than a Void, the fear evaporates. We realize that the work we do now—the shadow integration and the sharpening of our internal focus—is the necessary preparation for the next octave of existence..

The more integrated your signal is now, the more laminar (smooth) your transition into the steam phase will be. You aren't dying; you are finally stepping out of the limited 2% render and into the full resonance of the 98%.

Chapter 14: Endless Possibilities

Operating the 98% Lever

The Spectrum of the Miraculous

We often label things as "miraculous" or "impossible" simply because we don't understand the scale of the laws being applied. In a Scale Invariant universe, the same math that allows a subatomic particle to be in two places at once is the same math that allows a human being to maintain vibrant health or an ancient architect to move a 70-ton stone.

There is no difference in the physics of these events—only a difference in the magnitude of coherence. To a Cork living entirely in the 2%, these things look like magic. To a Soliton, they are simply the result of moving the lever in the 98% to influence the render in the 2%.

The Purposeful Storm: Health as Refinement

It is a common trap to assume that once you "unlock the code," you will enjoy a life of effortless ease. But the path of the Soliton is rarely a straight line. Many who walk this path experience profound health challenges or life-altering struggles.

These challenges are not glitches in your render; they are refinement protocols. From the perspective of the 98%, you may have purposefully dialed up the viscosity of your life—creating a health crisis or a financial collapse—to act as the friction necessary to sharpen your internal focus. You aren't fixing a broken body; you are using the struggle to find the resonance that eventually sustains it.

The Law of the Lever: Archimedes and the 98%

The ancient Greek mathematician Archimedes famously said: "Give me a place to stand, and a lever long enough, and I will move the world."

Most people spend their entire lives trying to "move the world" by pushing directly on the object. In our model, this is 2% vs. 2%. It is exhausting, linear, and limited by how much brute force you can apply. But a Soliton understands that the 98% Aether is the long end of the lever. When you find your "place to stand"—the Hub of your own consciousness—and you apply your focus to the 98% medium, you gain Non-linear leverage.

The Oscillator: The Multi-Dimensional Hand

If the Aether is the lever, the Heart is the hand that moves it. The heart is a multi-dimensional transducer, taking the high-pressure information of the 98% and stepping it down into the 2% physical render.

In spiritual circles, we often hear the instruction to "open the heart." Most treat this as a sentimental emotional state, but in the physics of the Soliton, an Open Heart is a state of zero impedance.

Impedance is the measure of a circuit's resistance to the flow of current. A "closed heart" is high-impedance—clamped down by the fears and traumas of the 2% ego, creating a hardware lock that prevents the signal of the 98% from flowing through. An "open heart" means the valve is fully open. You aren't just sending love; you are allowing the 98% to flow through the vessel without resistance. This makes your "hand on the lever" incredibly powerful, as it is backed by the full pressure of the universal medium.

The Viscosity of Belief: The Hidden Setup

Even with a strong hand on the lever, we must account for the coefficient of belief. In fluid dynamics, viscosity is the measure of a fluid's resistance to flow. In the 98% Aether, your belief systems act as the viscosity.

When you attempt to move your life in a new direction, but your emotional baseline is filled with doubt, you are essentially trying to move a lever through thick honey. Shifting your belief becomes necessary because it creates a clearer passage for your intention. As your inner world becomes less conflicted, your energy can move more freely, allowing your intent to flow smoothly into the render.

Recognizing that this internal alignment is a mechanical necessity changes everything. It shifts us away from hoping for a change and moves us into the seat of the operator. And while the language of wave mechanics and field dynamics might be modern, the understanding of this 98% lever is not new. In fact, some of the most profound instructions for this exact alignment were recorded over two millennia ago.

The Gnostic Secret: The Power Within

The word *Gnosis* does not mean "faith" or "belief"; it means Direct Knowledge—knowledge gained through the kind of witnessed experience we have been discussing throughout this book. Gnosticism was not a religion, but a technical inquiry into the nature of the "Matrix." The Gnostics were the original physicists of the spirit who realized that the material world was a render—a limited interface—and that the path to sovereignty was found by looking behind the paint to the source code of the Father (the Field).

When we look at Gnostic teachings today, we are not deviating from science; we are returning to a high-frequency understanding of information theory. They taught that the "Kingdom" was not a place in the clouds, but a state of resonance that could be achieved by integrating the self.

One of the most profound operator manuals for the 98% is found in the Gnostic Gospel of Thomas. In Saying 22, we find a technical description of what happens when a system achieves zero impedance:

"When you make the two into one, and when you make the inner as the outer, and the upper as the lower, and when

you make male and female into a single one... then you will enter [the kingdom]."

To the Sovereign being, "the kingdom" is not a destination; it is a state of resonance.

- The Inner as the Outer: This is the elimination of social camouflage. When your internal truth and your external broadcast are identical, you stop being a leaky system. Your signal-to-noise ratio reaches its maximum.
- The Upper as the Lower: This is the realization of Scale Invariance. Your biological cells (the lower) and your conscious intent (the upper) are finally phase-locked.
- The Two into One: This is the collapse of duality. You are no longer the fixer fighting the ocean; you have become the wave that *is* the ocean.

When you follow these instructions, you aren't praying for a result; you are stepping into the physics of the Soliton. A Soliton only exists when two opposing forces perfectly balance and integrate into a single, unbreakable phenomenon. When you finally make the "two into one," your consciousness achieves that exact state of self-sustaining propulsion. This is the moment the lever of the 98% becomes effortless to move.

The Hermetic Physics: The Original Source Code

The Gnostics understood the mechanics of the render perfectly, but they were actually drawing from a much older, foundational framework. Thousands of years before the Gospel of Thomas was written, the exact same physics were recorded in the ancient Hermetic tradition. The Hermetic principles and the Gnostic teachings are essentially describing the exact same thing; the Hermeticists were simply the original systems engineers who mapped the entire fluid dynamics of reality.

Throughout this book, we have explored the reality of Scale Invariance—the mechanical truth that the exact same physics apply at every level of magnification, from the microscopic cell to the cosmic field. The ancient Hermeticists recorded this exact same law. They didn't call it Scale Invariance; they called it the Principle of Correspondence: *"As Above, So Below."*

They realized that reality was not a series of random events, but a highly structured, responsive medium.

For thousands of years, history has recorded sovereign individuals who seemed capable of altering reality at will. But these were not supernatural deities operating above the law; they were simply operators who understood the mechanics of the 100%. They realized that the ancient Hermetic principles were not mystical poetry—they were a technical manual for wave mechanics.

When we translate the Seven Hermetic Principles through the lens of the Soliton, the entire physics engine of the render reveals itself:

- **1. The Principle of Mentalism (*"The All is Mind"*):** Everything we touch in the 2% physical world is just a "render" of the information organized in the 98% Field. The universe is a conscious, responsive medium.

- **2. The Principle of Correspondence (*"As Above, So Below"*):** This is the ancient definition of Scale Invariance. The exact same mechanical laws apply at every level of magnification, from the cellular biology of your body to the orbital mechanics of the cosmos.

- **3. The Principle of Vibration (*"Nothing rests; everything moves"*):** This is the foundational law of the Field. Every thought, emotion, and physical object is simply a frequency. To change the render of the 2%, you don't push the matter—you calibrate the oscillator.

- **4. The Principle of Polarity (*"Everything is dual"*):** Opposites are not enemies; they are just varying degrees of the exact same energy. This is the core physics of the Soliton. A Soliton actively uses the tension between two opposing forces to sustain itself. When you integrate these polarities rather than fighting them, you stop bouncing back and forth in a closed 360-degree loop. You turn that friction into the fuel for the +1 Engine, thrusting your consciousness upward into the next harmonic spiral.

- **5. The Principle of Rhythm (*"Everything flows, out and in"*):** This is the natural oscillation of existence. The tide goes out, and the tide comes in. A Soliton understands that this rhythmic movement is a permanent feature of the physical render. Rather than fighting the pendulum swing or being destroyed by its momentum, the Sovereign operator simply anticipates it—riding the wave naturally while remaining centered at the Hub.

- **6. The Principle of Cause and Effect (*"Every Cause has its Effect"*):** This is the physics behind what tradition calls "karma." No action exists in a vacuum. If you act from a place of friction or harm, you are writing destructive code into the 98% Field, which must eventually render as a destructive Effect in your 2% physical reality. The 98% is always the Cause; the 2% is always the Effect. The Sovereign operator doesn't try to cheat the system or fix the effects on the screen—they understand that their every output is a Cause, and they govern their actions accordingly.

- **7. The Principle of Gender (*"Everything has its Masculine and Feminine Principles"*):** This has nothing to do with biological sex; it is the fundamental physics of creation. It is the geometric balance between the "Masculine" (the electric outward thrust of intention) and the "Feminine" (the magnetic, receptive capacity to draw it in). Just as the Gnostics taught, making "the male and female into a single one" drops your impedance to zero, completing the circuit of the Soliton.

Chapter 15: The Frequency Bridge

The Individual as a Global Oscillator

The Mandate of the Field

Nikola Tesla famously observed that if you want to find the secrets of the universe, you must think in terms of energy, frequency, and vibration. He understood that the visible world is merely a persistent effect caused by an invisible cause.

For most of our history, we have been obsessed with the effect. We have tried to change the world by moving the furniture around in the 2% render. But as we enter the Age of Aquarius, the Master Code is becoming clear: To change the render, you must change the frequency of the broadcast. You are not an isolated "thing" sitting in a vacuum of empty space. You are an oscillator coupled to a high-pressure field of infinite information.

The Phase Change: From Pisces to Aquarius

As we transit the bridge between ages, the very nature of how we love and feel for one another is undergoing a fundamental phase change.

Love in the Age of Pisces (The Frequency of Sympathy): For the last 2,000 years, love was defined by a heavy, immersive focus on shared struggle. Compassion was often expressed through sympathetic resonance—the instinct to match the frequency of those in pain to show we care. While deeply human and well-intentioned, this meant we often fragmented our own signal to mirror the chaos of others. It was a high-impedance way of connecting; we essentially merged our waveforms with the suffering, which diluted our ability to provide a clear, restorative signal.

Love in the Age of Aquarius (The Frequency of Resonance): In this new era, love becomes a broadcast, not a transaction. It is the state of zero impedance. Here, compassion is the ability to witness the struggle of another without losing your own stability. You don't collapse your wave to match their pain; you hold your baseline so steadily that it provides a calming, rhythmic influence for them to tune into. You aren't "feeling their pain"; you are holding the frequency of their own wholeness.

The Friction of the Shift: The Great Squeeze

If this new age is about flow and resonance, why does the world look like it's tearing itself apart? To understand the current global chaos, we must look at the friction of the shift.

We are currently standing at the junction where two massive "tectonic plates" of time are rubbing against each other. The Plate of Pisces (defined by hierarchy, secrecy, and "ice" structures) is resisting the Plate of Aquarius (defined by sovereignty, transparency, and "fluid" resonance).

When these two ages rub together, they create mechanical heat. This heat manifests in our 2% render as geopolitical unrest, financial instability, and social friction. The old world is "high-impedance"—it is designed to resist the flow. As the 98% Aetheric pressure increases, the old, rigid pipes of the Piscean era are beginning to burst because they cannot handle the new, high-frequency "steam" of the Aquarian signal.

The chaos isn't the world ending; it is the world unlocking. The "ice" is being forced into a phase change, and that process is inherently turbulent. As a Soliton, your job isn't to stop the friction, but to remain fluid within it—to be the lubricant that allows the transition to happen with as much laminar flow as possible.

The Aquarian Render: Themes of the New Earth

As the Soliton wave of Aquarius moves through the collective, it creates a specific "laminar llow" that impacts every system on Earth. Because the 98% is now loading a more coherent setup, the old 2% structures are beginning to fracture.

- Sovereignty over Centralization: The Age of Pisces was the age of the "Intermediary." The Aquarian frequency is one of Radical Sovereignty. As individuals become their own frequency bridges, the need for centralized control evaporates.

- Transparency over Secrets: The new frequency acts as a global pressure wash, bringing hidden information to the surface. This is why the world feels so exposed; the global pipes are being cleared of thousands of years of un-integrated history.

- Unity through Diversity: The new world isn't about getting along through compromise; it's about Unified Polarity. It is the realization that we can all be unique, non-linear Solitons while still being part of the same ocean.

The Power of the Sovereign Presence

The greatest misconception of the old paradigm was that service required sacrifice. In the physics of Sovereignty, the opposite is true: The greatest service you can provide to the collective is your own coherence.

When you integrate your shadow and open your heart to a state of zero impedance, you become a frequency bridge. You create a low-resistance zone that allows the surrounding field to begin to organize itself. Like a tuning fork struck in a room full of silent ones, your resonance will naturally cause others to begin to vibrate in sympathy. You are lowering the viscosity of reality for everyone in your wake simply by refusing to be anything other than your sovereign self.

The Age of the Water Bearer

The symbol of Aquarius is the Water Bearer, often depicted as a figure pouring a never-ending stream from a vessel. This is the ultimate metaphor for the sovereign individual.

- The Vessel is your integrated 2% physical body.
- The Water is the high-pressure 98% Aetheric Truth.

The Water Bearer doesn't "hold" the water; they allow it to flow. They have reached a state where they can carry the immense pressure of the infinite without breaking.

Chapter 16: The 100% Perspective

The End of the Separation

The Final Reveal: The Seamless Whole

Throughout this journey, we have used the language of division to help the mind understand the mechanics of reality. We talked about the 2% Render and the 98% Aether. We talked about the Ice, the Water, and the Steam. We did this because, in a world of high viscosity, the mind needs 'things" to hold onto.

But as you stand here at the end of the code, it is time to let the final filter fall away. There is no 2% and there is no 98%. There is only the 100%.

The separation was a training wheel for your consciousness. You had to believe you were a separate Soliton so that you could learn to hold your own resonance. But the "mic drop" of the Sovereign path is the realization that the observer, the observed, and the medium of observation are a single, unbroken field. You aren't influencing the render from the outside; you are the render.

The Creator's Paradox

In the old world, we were told we were either victims of fate or manifestors of desires. Both are half-truths. A Sovereign Soliton understands the paradox: You are the wave, but you are also the ocean that allows the wave to exist.

When you look at your life—the struggles, the health challenges, the great squeeze of the global transition—you can finally see them for what they are: The Creator playing a game of Hide and Seek with itself. You purposefully dialed up the viscosity. You purposefully hid the code from yourself so that you could experience the sheer, brilliant triumph of finding it again.

The Physics of Grace: Living in the Flow

Now that the walls between the 2% and the 98% have dissolved, you enter the state of Grace. In the old paradigm, grace was something bestowed upon you by an external deity—a gift given from the outside. In the 100% perspective, Grace is a hydrodynamic state. It is the experience of living without resistance.

When you no longer fight the render, and when you no longer see challenges as enemies but as purposeful viscosity, the friction of life begins to evaporate. Life begins to align around you with a quiet, natural ease—not because you are forcing a specific outcome, but because you are finally available for the flow. You have cleared the pipes of your spiral memory to such a degree that the infinite pressure of the Aether can finally pour through you without causing turbulence.

The Sovereign Smile

There is a quiet, deep-seated peace that arises when the effort to be sovereign finally ceases and you simply realize that you have never been anything else. This is the Sovereign Smile.

It is the inner radiance you carry even when the 2% render appears turbulent—when the rigidity of the old world meets the fluid potential of the new. While the world may seem to struggle on the Rim, you are grounded in the Hub, witnessing the transition with the gentle certainty of one who understands the changing of the seasons. You aren't merely watching the physics; you are resting in the deep understanding of the cycle. You know that every shift is a necessary part of the unfolding, and that the 100% is already here, waiting only for our awareness to catch up to the flow.

The First Breath

This book was not written to give you new information. It was written to help you remember the frequency you already are. As you close these pages, the work doesn't start —the struggle simply ends.

You are a Water Bearer in an age that is thirsty for coherence. You don't need to shout, you don't need to convince, and you certainly don't need to save anyone. You simply need to breathe. With every coherent breath, you send a laminar pulse through the Aether that reminds the world that it is safe to melt.

The 100% is not a destination you reach. It is the realization that you never left.

Welcome to the Flow.

www.ingramcontent.com/pod-product-compliance
Lightning Source LLC
LaVergne TN
LVHW022057200926
848481LV00003B/964
* 9 7 9 8 9 9 3 6 0 6 5 1 4 *